AEROGEL AS A SUSTAINABLE CONSTRUCTION MATERIAL

Towards Net Zero Carbon Emissions

Steven Smith

Wisdom Publishers

To my loving wife, who has been my rock and my inspiration throughout this journey. Your unwavering support, encouragement, and patience have been invaluable to me. This book would not have been possible without you.

I also dedicate this book to my children, who bring me joy and laughter every day. You are my motivation to work hard and to make a positive impact on the world.

Lastly, I dedicate this book to all the passionate and dedicated individuals working tirelessly to promote sustainability in the construction industry. Your efforts are making a difference, and I hope this book will contribute to your important work.

Thank you all for your love, support, and inspiration.

The greatest threat to our planet is the belief that someone else will save it.

ROBERT SWAN

CONTENTS

INTRODUCTION

Sustainable construction has become a crucial challenge in both developed and developing countries over the past decade due to the negative impact of construction operations on the environment. Buildings generate 25% of all ozone-depleting chlorofluorocarbons emitted during the manufacture of building materials and building air conditioning. The emission of heat-trapping greenhouse gases has led to climate change, which is the most critical problem of the millennium due to its large environmental impact. Some greenhouse gases, particularly carbon dioxide, are abundant and have the ability to remain in the atmosphere for thousands of years. Buildings are important sources of CO_2 emissions. With its 2030 Agenda for Sustainable Development Goals, the United Nations had committed to preserving the ecosystem. One of the main goals of the UN Sustainable Development Goals is that different sectors and countries adopt practices that promote a healthy ecosystem and prevent ecosystem-damaging practices.

Buildings are the largest energy consuming sector in the world, accounting for over 40% of global energy consumption. In India alone, buildings consume 35% of all electricity generated, of which more than 50% is used for air conditioning. Achieving significant energy and emissions reductions in the building sector is a challenging but achievable policy goal. To keep global warming below 2 degrees Celsius, the IEA has advocated the need to reduce CO2 emissions in the construction sector by 77% by 2050. Due to the negative impact of the construction sector on the ecosystem, there is a growing call for sustainability practices in the industry. Several studies have attempted to improve sustainable practices in the construction sector. Sustainable construction management, life cycle analysis, corporate social responsibility, and recycling and waste reduction were explored to improve sustainable construction practices.

The construction sector is responsible for almost 45% of global greenhouse gas emissions, 11% of which result from the manufacture of building materials. Building material production contributes significantly to embodied energy, which further contributes to greenhouse gas emissions. One of the principles of sustainability is the choice of recyclable building materials. Alternative building materials have been extensively researched to promote sustainable construction and significantly reduce greenhouse gas emissions.

Sustainable building materials are a critical component of responsible construction practices that prioritize the health of the environment, human health, and the economy. These materials are designed to meet the needs of the present generation without compromising the ability of future generations to meet their own needs. One of the primary benefits of sustainable construction materials is their positive impact on the environment. Traditional construction practices are often associated with high levels of carbon emissions, which contribute to global climate change. Sustainable materials, on the other hand, are designed to reduce or eliminate the carbon footprint of construction projects by using renewable resources, minimizing waste, and reducing energy consumption. The energy saving capacity of sustainable building materials, among other advantages, makes them an important reference for sustainable development. The need for increased use of sustainable building materials has contributed to growing research in this area.

In order to promote sustainability with the use of alternative materials, phase change materials, alternative insulation materials, alternative cementitious materials, bamboo, sugar cane bagasse waste, alternative aggregates, etc have been recommended for use in construction operations. While these sustainable materials exist with more being added to them, this book is focused on aerogel as a sustainable construction material. Aerogel is a unique material that offers a wide range of benefits when used as a sustainable construction material. It is a solid material that is up to 99.8% air, which makes it incredibly lightweight while still being strong and durable. Aerogel has exceptional thermal insulation properties, making it an ideal material for use in buildings to reduce energy consumption and promote energy efficiency. It can reduce energy use by up to 50% and can help to reduce carbon emissions and energy costs. Aerogel also has excellent sound insulation properties. It can reduce noise levels by up to 100 decibels, making it ideal for use in buildings located in noisy environments or in sound-sensitive applications such as recording studios.

Aerogel is also fire-resistant, which is an important factor in building safety. Unlike other insulation materials, aerogel does not burn or produce toxic fumes when exposed to fire, making it a safer choice for buildings. It is made from silica or other natural materials and can be produced using renewable energy sources, which also makes it environmentally friendly. Aerogel is recyclable and can be reused in other applications, reducing waste and promoting a circular economy.

Divided into nine chapters, this book provides a detailed overview of aerogel, its unique properties, and its applications in the construction industry. It explains how aerogel can be used to promote energy efficiency and reduce carbon emissions, making it an essential tool for achieving net-zero carbon footprints in construction projects. The science behind aerogel, its production and manufacturing, and its use in different building applications are examined. It also presents case studies and real-world examples that showcase the potential of aerogel as a game-changing material in the pursuit of sustainable construction practices.

CHAPTER ONE:
AN OVERVIEW OF
SUSTAINABILITY

S ustainability is a term that is often used in discussions related to environmental issues, social responsibility, and economic growth. The concept of sustainability is rooted in the idea of meeting the needs of the present without compromising the ability of future generations to meet their own needs. The concept of sustainability is closely tied to the concept of sustainable development, which was first defined by the Brundtland Commission in 1987 as "development that meets the needs of the present without compromising the ability of future generations to meet their own needs." This definition emphasizes the importance of balancing economic growth, social well-being, and environmental protection. Sustainability can be viewed from several perspectives: environmental, economic, and social. From an environmental perspective, sustainability involves the responsible use and management of natural resources, the reduction of waste and pollution, and the preservation of biodiversity. From an economic perspective, sustainability involves ensuring that economic growth is balanced with social and environmental considerations, and that resources are used in a way that ensures their availability for future generations.

From a social perspective, sustainability involves ensuring that all members of society have access to basic needs such as food, water, shelter, and healthcare, and that their cultural and social identities are respected and protected.

The importance of sustainability cannot be overstated. In recent years, there has been growing recognition that our current patterns of consumption and production are not sustainable, and that urgent action is needed to address the environmental, economic, and social challenges we face. Climate change, loss of biodiversity, water scarcity, and social inequality are just a few of the many pressing issues that require a sustainable approach. To achieve sustainability, various strategies and practices are employed, such as the use of renewable energy sources, the implementation of sustainable agriculture and forestry practices, the reduction of waste and pollution, the promotion of sustainable transportation, and the adoption of sustainable urban planning and design. These strategies and practices aim to reduce the negative impacts of human activity on the environment, while promoting economic growth and social well-being.

One of the key strategies for achieving sustainability is the adoption of the circular economy model. The circular economy is an economic system that is based on the principles of reducing, reusing, and recycling materials and resources, in order to minimize waste and maximize the value of resources. This model aims to decouple economic growth from resource consumption, and to create a regenerative economic system that promotes sustainability. Another important strategy for achieving sustainability is the use of green technologies. Green technologies are technologies that are designed to reduce the negative environmental impact of human activity, while promoting economic growth and social well-being. Examples of green technologies include renewable energy sources such as solar and wind power, energy-efficient buildings and appliances, electric and hybrid vehicles, and sustainable agriculture and forestry practices.

The adoption of sustainable practices is not only important for the long-term health of the planet, but also for the long-term health of the economy and society. Sustainable practices can help to reduce costs, increase efficiency, and create new business opportunities. For example, the adoption of energy-efficient technologies can lead to significant cost savings for businesses and households, while the development of renewable energy sources can create new job opportunities and promote economic growth. As earlier indicated, the three pillars of sustainability are environmental, economic, and social considerations. Achieving sustainability requires a shift in the way we think about and use resources, and the adoption of strategies and practices that promote sustainable development. The adoption of a circular economy model and the use of green technologies are just two of the many strategies that can be employed to achieve sustainability. As we face the many challenges of the 21st century, it is clear that a sustainable approach is essential for the long-term health and well-being of the planet, the economy, and society. Through collective action and a commitment to sustainability, we can create a more just and equitable world for all, now and in the future.

Understand the history of sustainability

The concept of sustainability has been present throughout human history, although the term itself did not come into widespread use until the late 20th century. The idea of using resources in a responsible and sustainable manner can be traced back to ancient civilizations, such as the Greeks and Romans, who developed systems for managing water, soil, and other natural resources. In the Middle Ages, the concept of sustainability was closely tied to the idea of stewardship, or the responsible management of resources on behalf of future generations. This idea was reflected in the development of sustainable agriculture practices, such as crop rotation, which helped to maintain soil fertility and prevent soil erosion. The Industrial Revolution of the 18th and 19th centuries brought about significant changes in the way humans interacted with the environment. The rapid industrialization and urbanization that took place during this period led to increased pollution and degradation of natural resources. However, it also led to the emergence of new technologies and ideas that would pave the way for the modern sustainability movement.

In the early 20th century, a number of influential writers and thinkers began to explore the concept of sustainability in more detail. One of the most influential was Rachel Carson, whose book "Silent Spring" (1962) warned of the dangers of pesticides and other harmful chemicals to the environment and human health. This book helped to spark the modern environmental movement and increased public awareness of the need for sustainable practices. Another important figure in the history of sustainability was Buckminster Fuller, an architect and designer who developed the concept of "spaceship Earth." According to Fuller, the planet should be viewed as a closed system, with finite resources that must be managed in a sustainable manner in order to ensure the survival of future generations. In the 1960s and 1970s, sustainability became a more mainstream concept, as concerns about environmental degradation and resource depletion grew. The United Nations held its first conference on the environment in Stockholm in 1972, which helped to bring global attention to the issue of sustainability.

In the 1980s and 1990s, the concept of sustainable development emerged as a key framework for addressing environmental and social issues. The Brundtland Commission, established by the United Nations in 1983, defined sustainable development as "development that meets the needs of the present without compromising the ability of future generations to meet their own needs." This definition emphasized the importance of balancing economic, social, and environmental considerations. The Rio Earth Summit, held in 1992, was a key moment in the history of sustainability, as it brought together representatives from around the world to discuss global environmental issues and the need for sustainable development. The summit resulted in the adoption of Agenda 21, a comprehensive plan of action for sustainable development, as well as the creation of the United Nations Framework Convention on Climate Change.

Since the 1990s, sustainability has become an increasingly important issue for businesses and governments around the world. Many companies have adopted sustainable practices in order to reduce costs, increase efficiency, and promote long-term growth. Governments have also taken action to address environmental issues and promote sustainable development, through initiatives such as carbon pricing, renewable energy mandates, and green building codes. In recent years, the urgency of addressing environmental issues has become more pressing, as concerns about climate change, biodiversity loss, and other issues have increased. The adoption of the United Nations Sustainable Development Goals in 2015, which aim to address a wide range of environmental, social, and economic issues, represents a renewed global commitment to sustainability.

The concept of sustainability has a long and complex history, spanning thousands of years and numerous civilizations. While the term itself is a relatively recent addition to our vocabulary, the idea of using resources in a responsible and sustainable manner has been a key part of human culture and civilization for as long as humans have existed. Over time, the concept has evolved and become more sophisticated, reflecting the changing needs and challenges of society. Today, sustainability is a central issue for individuals, businesses, and governments around the world, as we work to create a more equitable and sustainable future for all. While the challenges we face are significant, the history of sustainability provides us with hope and inspiration, reminding us that we have the knowledge and tools we need to create a better world, if we are willing to take bold and decisive action.

Sustainability in the construction sector

The construction sector has a significant impact on the environment, contributing to a range of issues such as climate change, resource depletion, and environmental pollution. With the global population increasing, demand for housing and infrastructure is growing, leading to greater pressure on natural resources and ecosystems. However, the construction industry can play a significant role in promoting sustainability and reducing its impact on the environment. One of the main challenges in achieving sustainability in the construction sector is reducing the energy consumption and greenhouse gas emissions associated with buildings. Buildings are responsible for a significant proportion of global energy consumption and emissions, as they require heating, cooling, and lighting. To address this issue, various measures can be implemented, including the use of renewable energy sources, energy-efficient building materials and design, and green building standards.

Renewable energy sources such as solar and wind power are becoming increasingly popular in the construction sector. Solar panels, for example, can be integrated into building design, providing a sustainable source of energy for heating, cooling, and lighting. Wind turbines can also be incorporated into building design, particularly in high-rise buildings, to generate renewable energy. Energy-efficient building materials and design are another important way to promote sustainability in the construction sector. Passive design strategies, such as the use of natural ventilation and daylighting, can significantly reduce energy consumption and improve the comfort and livability of buildings. Insulation materials such as aerogel can also be used to reduce heat transfer and improve energy efficiency.

Green building standards such as LEED and BREEAM provide a framework for promoting sustainability in the construction sector. These standards set out criteria for sustainable building design, including energy and water efficiency, materials selection, and indoor environmental quality. By adopting these standards, the construction industry can help to promote more sustainable building practices and reduce the environmental impact of buildings. Sustainable building materials are also becoming increasingly important in the construction sector. Materials such as bamboo, straw bale, and rammed earth are renewable, low-impact, and biodegradable, making them an attractive option for environmentally conscious builders and designers. Traditional building materials such as wood and clay can also be sourced sustainably, reducing the environmental impact of construction.

Circular economy principles are also crucial for promoting sustainability in the construction sector. By adopting these principles, the industry can reduce waste and promote the reuse and recycling of materials. Modular and prefabricated building components, for example, can be easily disassembled and reused in other buildings, while the incorporation of recycled materials such as reclaimed wood and metal can significantly reduce the environmental impact of construction. Water management is another important aspect of sustainability in the construction sector. By implementing rainwater harvesting and greywater reuse systems, for example, buildings can reduce water consumption and promote more sustainable water management practices. Green roofs and other forms of green infrastructure can also help to reduce stormwater runoff and improve water quality in urban areas. Sustainable transportation and mobility strategies are also crucial for promoting sustainability in the construction sector. By incorporating bike lanes and pedestrian-friendly design into building and community planning, the industry can promote sustainable transportation practices and reduce greenhouse gas emissions. The adoption of electric and hybrid vehicles and the promotion of public transportation can also help to reduce emissions and improve air quality in urban areas.

Engaging with local communities and stakeholders is essential for promoting sustainability in the construction sector. By promoting sustainable building practices in local development projects, the industry can build trust and support for sustainable development initiatives. This can help to ensure that the benefits of sustainability are shared by all members of society, promoting a more equitable and sustainable future for all. Sustainability has become a crucial issue for the construction sector, and there are many ways in which the industry can promote more sustainable building practices. By adopting renewable energy sources, energy-efficient building materials and design, green building standards, circular economy principles, water management, sustainable transportation and mobility strategies, and engaging with local communities and stakeholders, the construction industry can significantly reduce its impact on the environment and promote a more sustainable future. With increasing demand for housing and infrastructure, it is essential that the construction sector adopts sustainable building practices to ensure that the environmental and social impacts of development are minimized. By promoting sustainability, the industry can build trust and support for sustainable development initiatives, ensuring that the benefits of sustainability are shared by all members of society.

CHAPTER TWO:
AEROGEL

A erogel is a highly porous material that is composed of 90% air and 10% solid material. It is sometimes referred to as "frozen smoke" due to its translucent appearance and extremely low density. Aerogel has a variety of unique properties, including low thermal conductivity, high surface area, and high porosity, which make it a valuable material for a wide range of applications.

There are several types of aerogel, including silica aerogel, carbon aerogel, and polymer aerogel. Silica aerogel is the most common type and is produced by removing the liquid from a silica gel under supercritical conditions. Carbon aerogel is produced by pyrolysis of a polymer aerogel, while polymer aerogels are produced by removing the liquid from a polymer gel. One of the most notable properties of aerogel is its low thermal conductivity, which is the ability of a material to conduct heat. Aerogel has a thermal conductivity that is 10 times lower than that of traditional insulation materials such as fiberglass or foam. This makes it an excellent insulator and has led to its use in a range of applications, including insulation for buildings, pipelines, and refrigeration systems. Another important property of aerogel is its high surface area, which is the amount of surface available for chemical reactions. Aerogel has a surface area that is several times higher than other materials, making it a valuable material for catalysis and filtration applications. For example, aerogel has been used as a catalyst support in the production of chemicals and as a filter for removing impurities from water and other liquids.

Aerogel also has a high porosity, which is the volume of space in a material that is not occupied by solid material. Aerogel has a porosity of up to 99%, which makes it an ideal material for absorbing sound and shock. This has led to its use in a range of applications, including acoustic insulation, shock absorbers, and protective coatings. Despite its unique properties, there are several challenges associated with the production of aerogel. The manufacturing process is complex and requires expensive equipment and specialized knowledge. Additionally, aerogel is a brittle material and can be prone to breaking if not handled carefully. However, advances in manufacturing techniques have led to the development of new types of aerogel with improved properties and lower production costs. For example, flexible aerogels have been developed that can be bent and compressed without breaking, opening up new possibilities for the use of aerogel in applications such as clothing and wearable technology. Aerogel has been a unique material with a wide range of applications in various industries. Its low thermal conductivity, high surface area, and high porosity make it an excellent insulator, catalyst support, and shock absorber. While there are challenges associated with its production and handling, advances in manufacturing techniques are making aerogel more accessible and affordable, opening up new possibilities for its use in the future.

History of Aerogel

Aerogel, also known as "frozen smoke," is a lightweight, highly porous material that is composed of 90% air and 10% solid material. It is renowned for its unique properties, including low thermal conductivity, high surface area, and high porosity, which make it a valuable material for a wide range of applications. The history of aerogel dates back to the early 20th century when it was first discovered by Samuel Kistler in the 1930s. Since then, the material has undergone significant developments, with many advancements being made in its manufacturing processes, properties, and applications. The discovery of aerogel can be traced back to Samuel Kistler's work at the College of the Pacific in California in the 1930s. Kistler was experimenting with the properties of gels and was fascinated by the gel's ability to retain its shape after being soaked in a liquid. He discovered that if he replaced the liquid in the gel with gas, he could create a solid material that was mostly air, yet retained the shape and size of the original gel. The resulting material had a very low density and was highly porous, making it an excellent insulator.

Despite its unique properties, the practical applications of aerogel were not immediately apparent, and it remained a curiosity in the scientific community for several decades. It wasn't until the 1980s that aerogel was produced commercially, with the development of new manufacturing techniques. The first commercially produced aerogels were made from silica, which is one of the most common materials used for aerogel production today. Silica aerogels are produced by removing the liquid from a silica gel under supercritical conditions, resulting in a solid material with a low density and high porosity. In the decades that followed, researchers continued to explore the properties of aerogel and developed new types of aerogels using different materials and manufacturing processes. For example, carbon aerogels were developed by pyrolysis of a polymer aerogel, while polymer aerogels were produced by removing the liquid from a polymer gel.

As the potential applications of aerogel became clearer, industries began to take notice. Aerogel's unique properties make it an excellent insulator, catalyst support, and shock absorber. It has been used in a range of applications, including insulation for buildings, pipelines, and refrigeration systems, catalysis in the production of chemicals, and as a filter for removing impurities from water and other liquids. Despite its many advantages, the production of aerogel is still a complex and expensive process, and the material is not yet widely used. However, advances in manufacturing techniques have led to the development of new types of aerogel with improved properties and lower production costs. For example, flexible aerogels have been developed that can be bent and compressed without breaking, opening up new possibilities for the use of aerogel in applications such as clothing and wearable technology.

In recent years, there has been a renewed interest in aerogel, particularly in the context of sustainable development. Aerogel's unique properties make it an excellent material for energy-efficient buildings and green infrastructure, and there is increasing research focused on finding ways to produce aerogel more sustainably. The history of aerogel is a story of discovery and innovation, with many researchers and industries contributing to its development. Despite the challenges associated with its production and handling, advances in manufacturing techniques are making aerogel more accessible and affordable, opening up new possibilities for its use in the future. As the demand for sustainable materials grows, aerogel is likely to play an increasingly important role in meeting these challenges.

Importance of aerogel to Sustainable Building Construction

Sustainable building construction has become increasingly important in recent years due to the urgent need to reduce the environmental impact of the built environment. One of the key areas of focus in sustainable building construction is improving the energy efficiency of buildings, which accounts for a significant portion of global energy consumption and greenhouse gas emissions. Aerogel, a highly porous, lightweight material with excellent insulating properties, is increasingly being recognized as an important material for achieving energy-efficient, sustainable buildings. One of the most significant benefits of using aerogel in building construction is its exceptional thermal insulation properties. Due to its high porosity and low thermal conductivity, aerogel can significantly reduce heat transfer through walls, floors, and roofs, leading to lower heating and cooling costs and improved energy efficiency. This is particularly important in regions with extreme temperatures, where buildings require significant heating and cooling to maintain comfortable indoor environments. Aerogel can also help to reduce the size and complexity of heating and cooling systems, further reducing the environmental impact of buildings.

Another important benefit of aerogel is its ability to provide acoustic insulation, reducing noise pollution from both inside and outside the building. This is especially important in urban areas, where noise pollution can be a significant problem and can negatively impact both the health and well-being of occupants and the environment. Aerogel's high surface area also makes it an excellent material for air filtration systems, removing impurities from the air and improving indoor air quality. Poor indoor air quality is a significant health concern, and can lead to respiratory issues, allergies, and other health problems. By using aerogel-based air filtration systems, building owners can ensure that the air inside their buildings is clean and healthy, promoting the health and well-being of occupants. Aerogel can also play an important role in sustainable building construction by reducing the environmental impact of building materials. Traditional insulation materials, such as fiberglass and polystyrene, are typically made from non-renewable resources and have a significant environmental footprint. Aerogel, on the other hand, can be produced using renewable materials and with low environmental impact, making it an excellent material for achieving sustainable building construction.

Using of aerogel in building construction can contribute to the overall sustainability of buildings by reducing the need for maintenance and replacement. Traditional insulation materials can degrade over time, reducing their effectiveness and requiring replacement. Aerogel, however, is a highly durable material that can last for the lifetime of a building, reducing the need for costly replacements and maintenance.

The use of aerogel in building construction can also contribute to the overall aesthetics of buildings. Traditional insulation materials can be bulky and unsightly, detracting from the overall appearance of a building. Aerogel, however, is a lightweight, thin material that can be easily integrated into building components such as walls, ceilings, and roofs, without detracting from the overall design and appearance of the building. Aerogel is an increasingly important material for sustainable building construction, with numerous benefits in terms of energy efficiency, acoustic insulation, air filtration, and environmental impact reduction. As the demand for sustainable building construction continues to grow, the use of aerogel is likely to become more widespread, helping to create buildings that are more energy-efficient, healthier, and more sustainable for both occupants and the environment.

Production of aerogel

The production process for aerogel is a complex and multi-step procedure that involves several stages to create the desired material. The steps involved in the production process are:

Selection of the precursor material: The first step in the production process is the selection of the precursor material. Commonly used precursor materials for aerogel include silica, alumina, and carbon. The choice of precursor material depends on the specific application requirements of the aerogel.

Sol-gel process: The next step is the sol-gel process, which involves the formation of a gel from the precursor material. This is achieved by dissolving the precursor material in a solvent, followed by adding a catalyst to promote the gelation process. The gelation process can take several hours to days, depending on the specific precursor material and solvent used.

Aging: After the gel is formed, it undergoes an aging process, which allows the gel to mature and gain strength. This process can take several days to weeks, depending on the specific gel and the aging conditions.

Drying: The gel is then subjected to a drying process to remove the solvent and create a solid material. This process can be done in several ways, including supercritical drying, freeze-drying, or ambient pressure drying. Supercritical drying is the most commonly used method, as it produces the highest quality aerogel.

Sintering: After drying, the aerogel is subjected to sintering, which involves heating the material to a high temperature to remove any remaining impurities and increase its mechanical strength. The sintering temperature and time are carefully controlled to ensure that the aerogel retains its unique properties and characteristics.

Surface modification: Finally, the aerogel may undergo surface modification to enhance its performance for specific applications. This can include adding functional groups to the surface or coating the surface with a thin layer of another material.

The production process for aerogel requires careful attention to detail to ensure the final product has the desired properties and characteristics. As technology advances and research continues, it is likely that more production methods will emerge to make the production of aerogel more efficient and cost-effective.

Aerogel as a Sustainable Building Material

Aerogel is a unique and innovative construction material that has a range of potential applications in sustainable building. Its composition and structure give it a number of properties that make it an attractive option for construction purposes, including its high porosity, low thermal conductivity, and lightweight nature. Aerogel is a highly porous material, with a structure that resembles a sponge or honeycomb. This high porosity gives it a large surface area, which makes it an excellent insulator against heat transfer. Its low thermal conductivity, which is several times lower than traditional insulation materials, means that it is highly efficient at reducing heat loss or gain through the building envelope, making it an ideal material for insulation purposes. This, in turn, can reduce energy consumption and lower greenhouse gas emissions associated with heating and cooling buildings. Another key advantage of aerogel is its light weight, which makes it easy to handle and transport. This makes it a suitable material for use in construction projects where weight is an important consideration, such as in the construction of high-rise buildings. Aerogel's light weight also means that it can be used in retrofitting applications, where adding additional weight to an existing structure may not be feasible.

Aerogel is also a durable material that can withstand a range of environmental conditions. Its high strength-to-weight ratio means that it can be used in structural applications, and its resistance to fire and moisture makes it suitable for use in a range of building applications. Additionally, its high optical clarity means that it can be used in glazing applications, allowing natural light to enter buildings while still providing excellent insulation properties. One of the most significant advantages of aerogel as a construction material is its environmental sustainability. Aerogel can be produced using renewable resources, and its long lifespan means that it has a lower environmental impact over the lifetime of the building than many other materials. Additionally, it is recyclable and can be reused in other applications at the end of its useful life.

Aerogel has many potential applications in sustainable building. Its lightweight, high porosity, low thermal conductivity, and environmental sustainability make it an attractive option for insulation, structural, glazing, and retrofitting applications. As technology continues to improve, it is likely that the use of aerogel in construction will become more widespread, further promoting sustainable building practices

CHAPTER THREE: PROPERTIES OF AEROGEL

The structure of aerogel is unique, with a three-dimensional network of nanoscale particles that form a highly porous solid. The particles are typically arranged in a random or amorphous structure, meaning that they lack long-range order or crystal structure. This lack of order gives aerogel its unique properties, including its high porosity, low density, and low thermal conductivity.

The composition of aerogel can be tailored to suit different applications. Silica aerogels are the most common type of aerogel, and they are produced by using a sol-gel process. In this process, a precursor solution of silica is mixed with a catalyst and a solvent. The solution is then allowed to gel, forming a solid network of particles. The solvent is then removed from the gel by a process called supercritical drying, in which the gel is exposed to high pressure and temperature conditions that cause the solvent to evaporate without disrupting the solid structure.

The resulting silica aerogel is a highly porous solid with a low density and high surface area. The pore size and distribution can be controlled by adjusting the composition and processing conditions. This control over the pore size and distribution allows aerogel to be tailored for different applications, such as thermal insulation, catalysis, and adsorption. Carbon aerogels are another type of aerogel that are composed of carbon nanoparticles. They are produced by using a similar sol-gel process, but with a carbon precursor instead of silica. Carbon aerogels have a lower density than silica aerogels and are highly conductive, making them useful for applications such as electrodes and energy storage.

Metal oxide aerogels are also possible, with materials such as titanium oxide and aluminum oxide being used as precursors. These aerogels have unique properties that make them useful for applications such as catalysis, gas sensing, and drug delivery. The composition and structure of aerogel are highly important in determining its properties and potential applications. By controlling the composition and processing conditions, it is possible to tailor aerogel for a wide range of uses, making it an increasingly important material for sustainable building construction and other applications.

Thermal Insulation

One of the most important properties of aerogel is its exceptional thermal insulation performance. Aerogel has the lowest thermal conductivity of any solid material, making it an ideal material for thermal insulation in buildings. Thermal conductivity is a measure of a material's ability to conduct heat. The lower the thermal conductivity, the better the insulation performance. Aerogel has a thermal conductivity of around 0.013 to 0.018 W/mK, which is several times lower than traditional insulation materials such as fiberglass, foam, and cellulose. Aerogel's thermal insulation properties arise from its unique structure and composition. The highly porous structure of aerogel, combined with its low density and small pore size, creates a material that is highly effective at trapping air and reducing heat transfer through conduction and convection.

In addition to its low thermal conductivity, aerogel also has a high melting point and is non-flammable, making it a safe and effective material for thermal insulation. It is also highly resistant to moisture and mold, further improving its performance and durability as a building insulation material. Aerogel can be used in a variety of building insulation applications, including walls, roofs, and floors. It can be applied as a spray, panel, or mat, making it easy to install and adaptable to different building configurations. Aerogel insulation can also be used in retrofit applications, allowing existing buildings to be upgraded with high-performance insulation without the need for major structural changes.

In addition to its thermal insulation properties, aerogel also has other benefits that make it a desirable material for sustainable building construction. These include its lightweight and durable nature, its ability to absorb sound and vibration, and its potential for use as a building material in its own right. Aerogel's exceptional thermal insulation properties make it a highly desirable material for sustainable building construction. Its unique composition and structure, combined with its safety, durability, and ease of installation, make it a valuable addition to the range of insulation materials available for use in building construction.

Sound Insulation

In addition to its exceptional thermal insulation properties, aerogel also has notable sound insulation capabilities, which can improve the acoustic performance of buildings. Sound insulation is the ability of a material to prevent sound transmission from one area to another. This is important in buildings, where sound can travel easily through walls, ceilings, and floors, causing noise pollution and disturbing occupants. Aerogel has a unique

STEVEN SMITH PH.D.

structure that allows it to absorb sound energy and dampen vibrations, making it an effective material for sound insulation. The high porosity and small pore size of aerogel allow sound waves to enter the material and dissipate as they encounter the solid network of interconnected particles. This results in a reduction in the amount of sound energy that is transmitted through the material.

Aerogel's sound insulation properties can be further improved by combining it with other materials, such as foam or fabric, to create composite materials that provide both thermal and acoustic insulation. These composites can be used in a variety of building applications, such as walls, ceilings, and floors, to provide improved sound insulation performance. Aerogel's sound insulation properties also make it a desirable material for use in high-performance acoustic panels and systems. These panels can be used to reduce noise pollution and improve the acoustic quality of buildings, such as in concert halls, theaters, and recording studios. Its sound insulation properties make it a valuable material for sustainable building construction. Its ability to reduce noise pollution and improve acoustic performance, combined with its thermal insulation properties and other benefits, make it a versatile material that can contribute to the creation of more comfortable and efficient buildings.

Fire Resistance

Fire resistance is an important property for any building material, as it plays a crucial role in protecting occupants and property from the spread of fire. Aerogel, as a material with unique properties, also exhibits some level of fire resistance. It is an inorganic material that is highly resistant to heat and flame. This is due to its high melting point and low thermal conductivity, which make it difficult for heat to transfer through the material. In addition, aerogel is non-combustible and does not produce any harmful gases or fumes when exposed to fire.

The specific fire resistance properties of aerogel depend on several factors, such as its composition, density, and thickness. Generally, higher-density aerogel with thicker layers will provide greater fire resistance than lower-density aerogel with thinner layers. Aerogel can also be combined with other materials, such as coatings or fabrics, to enhance its fire resistance properties. For example, aerogel can be used as a core material in fire-resistant panels, with a layer of fire-resistant fabric or coating applied to the surface for additional protection. In the construction industry, its fire resistance properties make it a valuable material for use in building insulation and other applications where fire safety is a concern. By using aerogel as a fire-resistant insulation material, builders and designers can help to reduce the spread of fire and improve the safety of buildings.

Its fire resistance properties can be further enhanced by combining it with other fire-resistant materials. As such, aerogel is a valuable material for sustainable building construction that can contribute to the creation of safer and more resilient buildings.

Mechanical Strength

Mechanical strength is an important property to consider in any construction material, as it determines the material's ability to withstand external forces and stresses. Aerogel, due to its unique composition and structure, exhibits some interesting mechanical properties. It is known for its high porosity, which gives it a low density and a high surface area-to-volume ratio. However, this porosity also means that aerogel is not inherently strong, and its mechanical properties can vary widely depending on its density, structure, and other factors. The mechanical strength of aerogel is typically measured in terms of its compressive strength, tensile strength, and shear strength. Compressive strength refers to the material's ability to resist deformation or collapse under a compressive load, while tensile strength refers to its ability to resist stretching or pulling apart. Shear strength, on the other hand, measures the material's resistance to forces that cause it to slide or twist.

Its mechanical strength can be improved by controlling its composition and structure during the production process. For example, the addition of reinforcing materials, such as fibers or particles, can help to enhance the material's mechanical properties. Additionally, altering the processing conditions, such as temperature and pressure, can also affect the resulting mechanical properties of the aerogel. In the construction industry, aerogel's mechanical properties make it a valuable material for use in a variety of applications. Its low density and high porosity make it a lightweight material that can be used to reduce the weight of building components without compromising on strength. For example, aerogel can be used as a core material in sandwich panels, with stronger materials like metals or polymers on the outer layers for added strength and durability. While aerogel is not inherently strong, its mechanical properties can be improved through careful control of its composition and structure during production. Its unique properties, such as low density and high porosity, make it a valuable material for use in lightweight construction applications where strength is also important.

Durability

Durability is an essential property in any construction material as it determines its lifespan and resistance to wear and tear. Aerogel, due to its unique composition and structure, exhibits some interesting durability properties. Aerogel is known for its excellent thermal insulation properties and its ability to resist moisture, which makes it resistant to many of the factors that can lead to material degradation. Its high porosity and low density also contribute to its durability, as they make it less susceptible to damage from impacts or vibrations. One factor that can affect the durability of aerogel is its exposure to UV radiation. Aerogel is known to be susceptible to photodegradation when exposed to UV radiation, which can cause it to lose some of its insulating properties over time. However, this can be mitigated by using coatings or encapsulation to protect the aerogel from UV radiation. Another factor that can affect the durability of aerogel is its resistance to mechanical stresses. As mentioned earlier, aerogel is not inherently strong, and its mechanical properties can vary depending on its composition and structure. Therefore, care must be taken to ensure that the aerogel is not subjected to mechanical stresses that could lead to deformation or damage.

Despite these challenges, aerogel has been shown to be a highly durable material when used appropriately in construction applications. Its unique properties make it well-suited for use in applications where thermal insulation and moisture resistance are important, such as in the insulation of pipes, roofs, and walls. Although aerogel can be susceptible to photodegradation and mechanical stresses, it can still be a highly durable material when used appropriately. Its unique properties make it well-suited for use in a variety of construction applications where thermal insulation and moisture resistance are important, and its durability can be enhanced through proper handling and protective measures.

CHAPTER FOUR: APPLICATIONS OF AEROGEL IN SUSTAINABLE BUILDING CONSTRUCTION

A erogel has several unique properties that make it an ideal material for insulation in sustainable building construction.

Insulation

As discussed earlier, it has very low thermal conductivity, making it an excellent insulator, and it is also lightweight, non-toxic, and non-flammable. These properties make it ideal for use in a wide range of applications where thermal insulation is required, including in building envelopes, walls, roofs, floors, and pipes. One of the most common applications of aerogel in sustainable building construction is in insulation. Aerogel can be used as a standalone insulation material or can be combined with other materials to create composite insulation materials with improved thermal performance. For example, aerogel can be combined with traditional insulation materials like fiberglass or cellulose to create high-performance insulation materials that offer better thermal insulation than traditional materials alone.

Aerogel insulation can be used in a variety of building applications, from residential homes to commercial buildings. In residential applications, it can be used in walls, roofs, and floors to improve the thermal performance of the building envelope, reducing energy consumption and lowering utility bills. In commercial applications, aerogel insulation can be used in pipes and ductwork to reduce heat loss and improve the efficiency of heating and cooling systems. Its insulation can also be used in retrofit applications to improve the energy efficiency of existing buildings. By adding aerogel insulation to the building envelope or upgrading the insulation in pipes and ductwork, building owners can reduce energy consumption, lower utility bills, and improve occupant comfort.

Wall Insulation

Aerogel insulation can also be used in wall insulation applications, both in new construction and retrofit projects. Walls are one of the most important components of a building's envelope, and their insulation is critical to reducing energy consumption and improving occupant comfort. Aerogel insulation can be installed in walls in a variety of ways, depending on the type of construction and the specific requirements of the project. In new construction, aerogel insulation can be installed in the wall cavity between the studs, or it can be applied directly to the exterior of the wall as a continuous insulation layer. In retrofit projects, aerogel insulation can be installed in the existing wall cavity, or it can be applied to the interior or exterior of the wall as a retrofit insulation layer. Its insulation can be used in both residential and commercial wall insulation applications. In residential applications, it can be used to insulate exterior walls, interior walls, and basement walls, while in commercial applications, it can be used to insulate walls in office buildings, hospitals, schools, and other types of buildings.

One of the main benefits of aerogel insulation in wall applications is its high thermal resistance. Because of its low thermal conductivity, aerogel insulation can provide a higher level of insulation in wall systems than traditional insulation materials. This can result in significant energy savings and lower heating and cooling costs for building owners. Another benefit of aerogel insulation in wall applications is its thin profile. Aerogel insulation can be much thinner than traditional insulation materials while still providing the same level of thermal resistance. This can be beneficial in applications where space is limited or where a thin profile is desired, such as in retrofit projects or in areas with limited wall thickness.

Roof Insulation

Aerogel insulation can also be used for roof insulation in sustainable building construction. Roofs are one of the most significant areas for heat loss in buildings, and a well-insulated roof can significantly reduce energy consumption and improve indoor comfort. There are several ways to install aerogel insulation for roof insulation applications, depending on the type of roof and the specific requirements of the project. In new construction, aerogel insulation can be installed between the roof rafters or applied directly to the roof deck as a continuous insulation layer. In retrofit projects, aerogel insulation can be installed on top of the existing roof as an additional insulation layer or installed in the roof cavity. One of the main benefits of aerogel insulation in roof applications is its high thermal resistance. Aerogel insulation has one of the highest R-values per inch of any insulation material on the market, meaning it provides superior thermal insulation performance compared to traditional insulation materials. This can result in significant energy savings and lower heating and cooling costs for building owners.

Another benefit of aerogel insulation in roof applications is its thin profile. Because of its high thermal resistance, aerogel insulation can be installed in thinner layers than traditional insulation materials while still providing the same level of thermal insulation performance. This can be beneficial in applications where space is limited or where a thin profile is desired, such as in retrofit projects or in areas with limited roof thickness. Aerogel insulation is also highly moisture-resistant, which is particularly important in roof applications. Moisture can significantly reduce the effectiveness of insulation materials, but aerogel insulation maintains its thermal insulation performance even in the presence of moisture. This can help prevent moisture-related problems, such as mold growth and structural damage, and prolong the lifespan of the building. Aerogel insulation is an excellent choice for roof insulation applications in sustainable building construction. Its high thermal resistance, thin profile, and moisture resistance properties make it a superior choice compared to traditional insulation materials, and it can help reduce energy consumption, lower heating and cooling costs, and improve indoor comfort in both residential and commercial buildings.

Floor Insulation

Aerogel insulation can also be used for floor insulation in sustainable building construction. Floors are another area of the building envelope where heat loss can occur, particularly in unheated or poorly insulated spaces such as basements and crawl spaces. A well-insulated floor can help reduce energy consumption and improve indoor comfort, particularly in colder climates. One of the main advantages of aerogel insulation in floor applications is its high thermal resistance. As mentioned earlier, aerogel insulation has one of the highest R-values per inch of any insulation material on the market, providing superior thermal insulation performance compared to traditional insulation materials. This can help reduce heat loss through the floor and improve energy efficiency, resulting in lower heating costs and a more comfortable indoor environment.

Aerogel insulation is also lightweight and thin, making it ideal for floor insulation applications. Its thin profile means that it can be installed in areas with limited vertical clearance, such as crawl spaces and underfloor cavities, without significantly impacting the height of the finished floor. This can be particularly beneficial in retrofit projects, where it may be challenging to add insulation to existing floors without significantly altering the height of the finished space. Another benefit of aerogel insulation in floor applications is its durability. Aerogel insulation is highly resistant to compression and can withstand heavy loads without losing its thermal insulation performance. This is important in floor applications, where the insulation material may be subjected to significant weight and pressure, particularly in commercial or industrial buildings.

Aerogel insulation is also resistant to moisture, which is important in floor applications. Moisture can lead to mold growth, rot, and other problems that can affect the structural integrity of the building and reduce the effectiveness of the insulation material. Aerogel insulation is highly resistant to moisture and can maintain its thermal insulation performance even in wet or humid conditions, helping to protect the building and prolong its lifespan. Aerogel insulation is an excellent choice for floor insulation applications in sustainable building construction. Its high thermal resistance, thin profile, durability, and moisture resistance properties make it a superior choice compared to traditional insulation materials. It can help reduce energy consumption, lower heating costs, and improve indoor comfort in both residential and commercial buildings, particularly in colder climates or in spaces with limited vertical clearance.

Glazing

In construction, glazing refers to the use of glass or other transparent or translucent materials to create windows, skylights, glass walls, or other features that allow natural light into a space. Glazing is an important aspect of building design, as it can affect the energy efficiency, comfort, and aesthetic appeal of a building. Glazing systems can be made from a variety of materials, including glass, acrylic, polycarbonate, and even fabric. They can also vary in terms of their energy efficiency, with some glazing systems designed to reduce heat transfer and improve insulation. One common type of glazing used in construction is insulated glass, which consists of two or more layers of glass separated by a sealed air space. This type of glazing can help reduce heat transfer and improve energy efficiency. Another important consideration when designing and installing glazing systems is safety. Glass can be fragile and pose a safety risk if it breaks or shatters, so it is important to use safety glass or other safety features, such as tempered glass, laminated glass, or safety film, to reduce the risk of injury.

Window Glazing

One of the most significant uses of aerogel is for window glazing. When used in window glazing, aerogel can significantly reduce the transfer of heat through windows, resulting in improved energy efficiency and reduced heating and cooling costs. This is especially important in buildings where energy consumption is a significant concern. Aerogel can be used in window glazing as a transparent insulation material between two layers of glass in an insulated glass unit. The aerogel material is highly effective at reducing the amount of heat that passes through the glass, while still allowing natural light to pass through.

Another application of aerogel in window glazing is to use it as a coating on the surface of the glass. This coating can reflect infrared radiation, which helps to reduce heat transfer through the glass, while still allowing visible light to pass through. This can result in significant energy savings in buildings, as the need for heating and cooling is reduced. The use of aerogel in window glazing is a sustainable building practice that can help to improve the energy efficiency and comfort of buildings while also providing a clear and transparent glazing material. It is a promising technology that has the potential to make a significant contribution to sustainable building practices.

Skylight Glazing

Aerogel can be applied for skylight glazing in building construction in two main ways: as an insulation material between two layers of glass or as a coating on the surface of the glass. When used as an insulation material, aerogel can be placed between two layers of glass to create an insulated glass unit. This allows for maximum insulation properties while still allowing natural light to pass through. The aerogel material is highly effective at reducing the amount of heat that passes through the glass, resulting in improved energy efficiency and reduced heating and cooling costs. The use of aerogel in this way can also reduce condensation and help to maintain a comfortable indoor environment.

Alternatively, aerogel can be used as a coating on the surface of the glass. This coating can reflect infrared radiation, which helps to reduce heat transfer through the skylight, while still allowing visible light to pass through. This can result in significant energy savings in buildings, as the need for heating and cooling is reduced. Aerogel can be applied to the glass in a variety of ways, including spray application or as a film that is laminated to the surface of the glass. The thickness and application method of the aerogel coating can be adjusted to suit the specific requirements of the building and skylight design. The use of aerogel in skylight glazing is a sustainable building practice that can help to improve the energy efficiency and comfort of buildings while also providing a clear and transparent glazing material. It is a promising technology that has the potential to make a significant contribution to sustainable building practices.

Soundproofing

Aerogel can be used in building construction for soundproofing purposes in a variety of ways. The unique properties of aerogel, including its high porosity and low density, make it an effective material for reducing sound transmission. One way that aerogel can be used for soundproofing is as an insulation material within walls, floors, and ceilings. When used in this way, aerogel can help to reduce sound transmission by absorbing sound waves and reducing the amount of vibration that is transmitted through the building structure.

Another way that aerogel can be used for soundproofing is as a coating on the surface of walls or ceilings. The aerogel coating can act as a barrier to sound waves, preventing them from passing through the building structure. This can be particularly effective in buildings where noise pollution is a concern, such as those located near busy roads or airports. In addition, aerogel can be used in building construction for soundproofing in the form of acoustic panels. These panels are typically made from a combination of aerogel and other materials such as fiberglass or foam. The panels can be installed on walls or ceilings to absorb sound waves and reduce noise transmission. This approach can be particularly effective in buildings such as recording studios, concert halls, and theaters, where high-quality acoustics are essential.

The use of aerogel in building construction for soundproofing purposes is a promising technology that has the potential to make a significant contribution to sustainable building practices. By reducing sound transmission, buildings can be made more comfortable and livable, improving the well-being of occupants and reducing the negative impacts of noise pollution on the environment.

Fireproofing

Aerogel can be used in building construction for fireproofing purposes in a variety of ways. The unique properties of aerogel, including its high porosity, low density, and high-temperature resistance, make it an effective material for reducing fire risk and improving fire safety in buildings. One way that aerogel can be used for fireproofing is as an insulation material within walls, floors, and ceilings. When used in this way, aerogel can help to reduce the spread of flames and heat in the event of a fire. Aerogel insulation can be used to fill the spaces between walls, for example, creating a fire barrier that can help to prevent flames from spreading from room to room. In addition, aerogel insulation can help to prevent the transfer of heat, reducing the risk of structural damage and improving the safety of occupants.

Another way that aerogel can be used for fireproofing is as a coating on the surface of building materials. Aerogel coatings can be applied to walls, floors, and ceilings to create a fire-resistant barrier that can help to prevent the spread of flames and heat. Aerogel coatings can also be applied to electrical cables and other components to prevent them from overheating and causing fires. Aerogel can be used in building construction for fireproofing in the form of fire-resistant boards or panels. These boards and panels are typically made from a combination of aerogel and other fire-resistant materials such as cement or fiberglass. They can be used to create fire-resistant barriers or to line walls and ceilings, improving the safety of buildings and their occupants in the event of a fire. The use of aerogel in building construction for fireproofing purposes is a promising technology that has the potential to make a significant contribution to sustainable building practices. By reducing the risk of fire and improving fire safety in buildings, aerogel can help to protect the environment and the well-being of building occupants.

Facade Systems

Aerogel can be used in building construction for facade systems, which are the exterior surfaces of buildings that are designed to provide a protective and decorative layer. The unique properties of aerogel, including its high porosity, low density, and high thermal insulation properties, make it an ideal material for use in facade systems. One way that aerogel can be used in facade systems is as an insulation material. Aerogel insulation can be used to improve the energy efficiency of buildings by reducing heat loss through the facade. This can help to reduce heating and cooling costs and improve the comfort of building occupants. Aerogel insulation can be applied as a coating or as a panel system, and can be integrated with other facade components such as cladding, windows, and doors.

Another way that aerogel can be used in facade systems is as a component of translucent or transparent facades. Aerogel can be used to create lightweight, high-performance glazing systems that offer superior thermal insulation properties while allowing natural light to enter the building. These types of facades are particularly useful in buildings such as offices, schools, and hospitals, where natural light is important for occupant well-being and productivity. Aerogel can be used in facade systems as a decorative element. Aerogel panels can be used to create intricate patterns or designs on the exterior of buildings, adding visual interest and texture to the facade. This approach can be particularly effective in buildings such as museums, galleries, and cultural centers, where the facade is an important part of the building's identity.

Using aerogel in building construction for facade systems is a promising technology that has the potential to make a significant contribution to sustainable building practices. By improving energy efficiency, enhancing natural light, and adding aesthetic value to buildings, aerogel can help to create more sustainable and livable environments for building occupants.

Energy Efficiency

Aerogel can be used in building construction to improve energy efficiency by reducing heat loss or gain through walls, floors, roofs, and windows. The unique properties of aerogel, including its high porosity, low density, and high thermal insulation properties, make it an ideal material for improving energy efficiency in buildings. One way that aerogel can be used for energy efficiency in buildings is as insulation material. Aerogel insulation can be applied to the interior or exterior walls, floors, roofs, or ceilings of buildings to reduce heat loss or gain. This can help to reduce heating and cooling costs and improve the comfort of building occupants. Aerogel insulation can be integrated with other building systems, such as HVAC (heating, ventilation, and air conditioning) systems and renewable energy systems, to further improve energy efficiency. Another way that aerogel can be used for energy efficiency in buildings is as a component of windows. Aerogel can be used to create high-performance windows that provide superior thermal insulation properties while allowing natural light to enter the building. These types of windows are particularly useful in buildings such as offices, schools, and hospitals, where natural light is important for occupant well-being and productivity. By reducing the need for artificial lighting and heating or cooling, aerogel windows can help to reduce energy consumption and improve energy efficiency in buildings.

Aerogel can be used for energy efficiency in buildings by creating lightweight, high-performance building envelopes. These envelopes can be used to reduce the overall weight of the building, which can help to reduce the amount of energy needed to transport building materials and construct the building. This approach can be particularly effective in buildings that are located in remote or difficult-to-access areas, where transportation costs are high. Using aerogel in building construction for energy efficiency is a promising technology that has the potential to make a significant contribution to sustainable building practices. By reducing energy consumption, improving occupant comfort, and reducing greenhouse gas emissions, aerogel can help to create more sustainable and energy-efficient buildings.

CHAPTER FIVE: ADVANTAGES OF AEROGEL AS A SUSTAINABLE BUILDING MATERIAL

A erogel has several advantages as a sustainable building material that can help to improve energy efficiency in buildings:

Energy Efficiency

High Thermal Insulation: Aerogel has a high thermal insulation property that can help to reduce heat transfer through the building envelope. This can help to reduce heating and cooling costs and improve the energy efficiency of buildings.

Lightweight: Aerogel is a lightweight material, which can help to reduce the weight of the building envelope. This can help to reduce the amount of energy required for transportation and construction of the building.

Durable: Aerogel is a durable material that can withstand harsh environmental conditions, such as extreme temperatures, moisture, and UV radiation. This can help to extend the lifespan of the building envelope and reduce maintenance costs.

Versatile: Aerogel can be used in a variety of building applications, such as insulation, windows, skylights, and facade systems. This versatility can help to reduce the need for multiple materials and simplify the construction process.

Environmentally Friendly: Aerogel is a sustainable building material that is made from silica, a natural and abundant resource. It is also recyclable and can be reused in other building applications, reducing waste and promoting a circular economy.

The use of aerogel as a sustainable building material can help to improve energy efficiency in buildings, reduce greenhouse gas emissions, and promote sustainable building practices. Its high thermal insulation property, lightweight, durability, versatility, and environmental friendliness make it a promising material for the construction of energy-efficient and sustainable buildings.

Reduced Carbon Footprint

Aerogel has several advantages as a sustainable building material that can help to reduce the carbon footprint of buildings:

Energy Efficiency: As mentioned earlier, aerogel has high thermal insulation properties, which can help to reduce the heat transfer through the building envelope. This can help to reduce the energy consumption for heating and cooling, resulting in lower greenhouse gas emissions.

Reduced Material Consumption: Aerogel is a lightweight material, which means that less material is needed to achieve the same level of insulation as traditional insulation materials. This can help to reduce the amount of raw materials required for construction and transportation, resulting in lower carbon emissions.

Recyclable: Aerogel is a recyclable material, which means that it can be reused in other building applications, reducing waste and promoting a circular economy. This reduces the need for new materials and the associated carbon emissions from production and transportation.

Durability: Aerogel is a durable material that can withstand harsh environmental conditions, such as extreme temperatures, moisture, and UV radiation. This can help to extend the lifespan of the building envelope and reduce maintenance costs, resulting in lower carbon emissions.

Renewable Raw Material: The raw material used to make aerogel is silica, which is a natural and abundant resource. This means that it is a renewable resource that can be sustainably sourced, reducing the impact on the environment and promoting a more sustainable building industry.

Aerogel as a sustainable building material can help to reduce the carbon footprint of buildings by reducing energy consumption, raw material consumption, waste, and maintenance requirements. Its recyclability, durability, and use of renewable raw materials make it a promising material for the construction of sustainable buildings with a reduced carbon footprint.

Long-Term Cost Savings

Here are the advantages of aerogel as a sustainable building material for long-term cost savings:

Energy Efficiency: Aerogel has a high thermal insulation property, which can help to reduce the energy consumption required for heating and cooling the building. This can result in significant long-term cost savings on energy bills.

Durability: Aerogel is a durable material that can withstand harsh environmental conditions, such as extreme temperatures, moisture, and UV radiation. This can help to extend the lifespan of the building envelope, reducing maintenance costs and avoiding the need for frequent repairs or replacements.

Lightweight: Aerogel is a lightweight material, which can reduce transportation and installation costs, as well as the structural requirements of the building envelope.

Versatility: Aerogel can be used in a variety of building applications, such as insulation, windows, skylights, and facade systems. This versatility can help to reduce the need for multiple materials and simplify the construction process, leading to cost savings.

Recyclable: Aerogel is a recyclable material, which can be reused in other building applications, reducing waste and promoting a circular economy. This can reduce the need for new materials and the associated costs of production and transportation.

Health and Safety Benefits

Aerogel also has several advantages as a sustainable building material that can provide health and safety benefits:

Non-Toxic: Aerogel is non-toxic and does not contain any harmful chemicals or gases, making it safe for use in indoor environments.

Fire Resistance: Aerogel has excellent fire-resistant properties and can withstand high temperatures without releasing toxic gases or fumes. This can help to prevent the spread of fires and protect the occupants of the building.

Moisture Resistance: Aerogel is highly resistant to moisture, which can help to prevent the growth of mold and mildew, reducing the risk of respiratory problems.

Noise Reduction: Aerogel has sound-absorbing properties, which can help to reduce noise levels inside the building. This can improve the comfort and well-being of the occupants.

Improved Indoor Air Quality: Aerogel can help to improve indoor air quality by reducing the infiltration of outdoor pollutants and allergens. This can be particularly beneficial for people with respiratory problems or allergies.

CHAPTER SIX: CHALLENGES AND LIMITATIONS OF USING AEROGEL IN SUSTAINABLE BUILDING CONSTRUCTION

While the use of aerogel in sustainable building construction offers many advantages, there are also several challenges and limitations to its widespread adoption.

Cost

One significant challenge is the cost of aerogel, which can be a barrier to its use in some building applications. This section will explore the cost challenges and limitations of using aerogel in sustainable building construction. Aerogel is a high-performance material that has many unique properties, including low density, high porosity, and excellent thermal insulation. These properties make it an attractive material for use in building construction, particularly in applications such as insulation, windows, and facade systems. However, the cost of producing and using aerogel can be significantly higher than other traditional building materials, which can limit its adoption.

The cost of aerogel is driven by several factors, including the cost of raw materials, production processes, and transportation. Aerogel is made from silica, a common mineral that is abundant in the earth's crust. However, the production process for aerogel is complex and requires specialized equipment, which can add to the cost. Additionally, the transportation of aerogel can be expensive due to its low density and fragile nature.

One of the main cost limitations of using aerogel in building construction is the high initial cost of installation. Compared to traditional insulation materials, aerogel can be up to 10 times more expensive per square foot. This cost difference can make it difficult for builders to justify the use of aerogel, particularly for large projects.

Another limitation of using aerogel in building construction is the limited availability of the material. While the demand for aerogel is growing, it is still a relatively niche market, and there are few suppliers that produce it. This limited supply can drive up the cost of aerogel and limit its adoption in some regions.

Despite the high cost of aerogel, there are several ways that builders and designers can work to reduce costs and make it a more feasible option for sustainable building construction. One approach is to use aerogel in combination with other building materials, such as concrete or steel, to reduce the overall amount of aerogel required. Additionally, improvements in production technology and increased competition among suppliers could help to drive down the cost of aerogel over time.

Another approach to reducing the cost of aerogel is to consider its long-term cost savings. While the initial installation cost of aerogel may be higher than traditional insulation materials, the long-term cost savings from improved energy efficiency and reduced maintenance costs can offset the initial investment. By considering the lifecycle cost of the building, builders and designers can make a more informed decision about the use of aerogel in sustainable building construction. Using aerogel in sustainable building construction offers many advantages, including improved energy efficiency, durability, and safety. However, the high cost of aerogel can be a significant barrier to its widespread adoption. Builders and designers must carefully consider the cost limitations and challenges of using aerogel and work to find ways to reduce costs and make it a more feasible option for sustainable building construction.

Availability

One of the challenges and limitations of using aerogel in sustainable building construction is the limited availability of the material. Despite its many advantages, the use of aerogel in building construction is still a relatively niche market, and there are few suppliers that produce it. Aerogel is a high-performance material that has many unique properties, including low density, high porosity, and excellent thermal insulation. These properties make it an attractive material for use in building construction, particularly in applications such as insulation, windows, and facade systems. However, the limited availability of aerogel can make it challenging to use in building projects.

There are several reasons why aerogel is not yet widely available in the construction industry. One reason is the complexity of the production process. Aerogel is made from silica, a common mineral that is abundant in the earth's crust. However, the production process for aerogel is complex and requires specialized equipment and expertise. This complexity can make it challenging for new suppliers to enter the market and can limit the supply of aerogel.

Another reason for the limited availability of aerogel is the relatively low demand for the material. While the demand for aerogel is growing, it is still a relatively new material in the construction industry, and many builders and designers are not yet familiar with its properties and applications. This limited demand can make it difficult for suppliers to justify the investment required to produce and supply aerogel.

The limited availability of aerogel can have several implications for sustainable building construction. One implication is that the cost of aerogel can be higher than other traditional building materials. The limited supply of aerogel can drive up the cost of the material, making it difficult for builders and designers to justify its use in some applications. Additionally, the limited availability of aerogel can make it difficult to source the material for large building projects, particularly in regions where there are few suppliers.

Another implication of the limited availability of aerogel is that it can limit the development of new applications for the material. Builders and designers are often hesitant to use a new material unless it is widely available and has a proven track record of performance. The limited availability of aerogel can make it difficult for builders and designers to experiment with the material in new applications, which can limit its adoption and slow its development as a sustainable building material.

Despite the challenges and limitations of the limited availability of aerogel, there are several ways that builders and designers can work to overcome these obstacles. One approach is to partner with suppliers and manufacturers to help them develop and scale up their production of aerogel. Builders and designers can provide feedback on the performance of aerogel in building applications and help suppliers identify new applications for the material.

Another approach to overcoming the limited availability of aerogel is to collaborate with other builders and designers to increase demand for the material. By working together, builders and designers can help to create a larger market for aerogel and drive down the cost of the material. Additionally, increased demand for aerogel can incentivize suppliers to invest in the production and supply of the material, which can help to increase the availability of aerogel in the market. The limited availability of aerogel is a challenge and limitation to its widespread adoption as a sustainable building material. The complexity of the production process, low demand, and limited supply of the material can make it difficult to use in building projects. However, by partnering with suppliers, collaborating with other builders and designers, and increasing demand for the material, builders and designers can work to overcome these obstacles and make aerogel a more widely available and accessible material for sustainable building construction.

Installation and Handling

Aerogel is a highly effective insulation material with several advantages for sustainable building construction. However, there are also some challenges and limitations that must be considered when using aerogel in building projects. One of the main challenges is the installation and handling of aerogel, which can be complex and time-consuming. Aerogel is a delicate and brittle material that can easily break and crumble if mishandled. It also has a tendency to absorb moisture, which can cause it to lose its insulating properties. Therefore, it is crucial to take special care when handling and installing aerogel in buildings.

One of the primary challenges with the installation of aerogel is the need for specialized equipment and expertise. Because aerogel is so fragile, it must be installed using special techniques and tools, such as vacuum-sealed bags, to prevent damage during transport and installation. Additionally, the application process can be time-consuming and labor-intensive, adding to the overall cost of the project.

Another challenge with the installation of aerogel is the need for careful attention to detail during the process. This is because aerogel must be installed in a way that ensures there are no gaps or spaces between the material and the building envelope. Even small gaps can significantly reduce the insulating effectiveness of the material, which can compromise the energy efficiency of the building.

The installation of aerogel can also pose challenges when it comes to retrofitting existing buildings. This is because the material is typically applied in layers, which can add thickness to the walls or roof, potentially causing issues with the building's structural integrity or interfering with existing features like doors or windows.

In addition to the challenges with installation and handling, there are also limitations to the availability of aerogel in sustainable building construction. While the demand for this material is growing, it is still relatively new and not widely available. This can make it difficult to source the material, especially in remote or rural areas.

Furthermore, the cost of aerogel can be a limitation for some building projects. Although aerogel has long-term cost savings potential due to its high insulation performance, it is currently more expensive than traditional insulation materials like fiberglass or foam. This can make it cost-prohibitive for some building projects, particularly those with tight budgets or limited funding.

Another limitation of using aerogel in sustainable building construction is the need for specialized expertise. This is because the installation process requires specialized knowledge and skills to ensure that the material is installed correctly and does not compromise the integrity of the building envelope. Without proper expertise, there is a risk that the insulation may be improperly installed, leading to issues with energy efficiency, moisture management, and structural integrity.

Finally, another limitation of aerogel is its limited range of applications. While aerogel is highly effective as insulation and has several other benefits for sustainable building construction, it may not be suitable for all building applications. For example, the fragility of the material and its thickness requirements may make it challenging to use in certain applications, such as in wall cavities or other confined spaces.

While aerogel has several advantages as a sustainable building material, there are also challenges and limitations that must be considered. The installation and handling of aerogel require specialized expertise, tools, and techniques, which can add to the overall cost and complexity of the building project. Additionally, the availability and cost of aerogel may be limiting factors for some building projects. Despite these challenges, however, the benefits of using aerogel in sustainable building construction are significant, making it a valuable material for building professionals to consider.

CHAPTER SEVEN: CASE STUDIES OF AEROGEL APPLICATIONS IN SUSTAINABLE BUILDING CONSTRUCTION

NASA's Sustainability Base

NASA's Sustainability Base, located at the NASA Ames Research Center in California, is a prime example of the successful application of aerogel in sustainable building construction. The Sustainability Base is a LEED-Platinum certified building, which means it meets the highest standards of sustainable design and operation. It was designed by William McDonough + Partners and completed in 2011.

Aerogel was used in the Sustainability Base as insulation for the building envelope, which includes the walls, roof, and foundation. The insulation was critical to the building's energy efficiency, which was a top priority for the project. The building was designed to be net-zero energy, meaning it produces as much energy as it consumes over the course of a year. The aerogel insulation used in the Sustainability Base is made by Aspen Aerogels, a company that specializes in aerogel insulation products. The product used in the Sustainability Base is called Spaceloft, which is a flexible aerogel blanket that can be easily installed in walls, roofs, and floors.

According to Aspen Aerogels, Spaceloft has a thermal conductivity of 14 milliwatts per meter Kelvin (mW/mK), which is about four times less than the thermal conductivity of traditional fiberglass insulation. This means that Spaceloft provides better insulation performance than fiberglass, while using less material. The Sustainability Base also features other sustainable design strategies, such as a green roof, natural ventilation, and a photovoltaic system. The combination of these strategies, along with the aerogel insulation, has allowed the building to achieve its net-zero energy goal.

The Sustainability Base has been recognized with several awards for its sustainable design, including the 2012 Green GOOD DESIGN Award, the 2012 National AIA COTE Top Ten Green Projects Award, and the 2013 ASHRAE Technology Award.

Another case study of aerogel application in sustainable building construction is the use of aerogel-filled polycarbonate panels in the restoration of the Chicago Union Station Great Hall. The restoration was completed in 2018, and the aerogel-filled panels were used to replace the original single-pane glass panels in the skylights. The aerogel-filled polycarbonate panels were chosen for their insulation performance, as well as their ability to filter out ultraviolet (UV) radiation. The panels were manufactured by Cabot Corporation, a company that specializes in specialty chemicals and performance materials. According to Cabot Corporation, the aerogel-filled panels have a thermal conductivity of 0.7 W/mK, which is about three times less than the thermal conductivity of traditional double-pane glass. This means that the panels provide better insulation performance than glass, while still allowing natural light to pass through.

The aerogel-filled panels were also chosen for their ability to filter out UV radiation, which can cause damage to artwork and other materials in the Great Hall. The panels filter out up to 99% of UV radiation, providing protection for the historic space and its contents. The use of aerogel-filled panels in the restoration of the Chicago Union Station Great Hall has been recognized with several awards, including the 2019 National AIA COTE Top Ten Green Projects Award and the 2019 AIA Chicago Distinguished Building Honor Award. In both of these case studies, aerogel has been successfully applied in sustainable building construction to achieve energy efficiency and other sustainable design goals. However, the challenges and limitations of using aerogel, such as cost and availability, must also be considered when evaluating its potential applications.

The Net-Zero Energy Research House

The Net-Zero Energy Research House is a groundbreaking project in sustainable building construction that demonstrates how innovative technologies and designs can be used to achieve net-zero energy consumption in residential buildings. The project was developed by the National Research Council of Canada and is located in Ottawa, Ontario. The Net-Zero Energy Research House was designed to achieve net-zero energy consumption by incorporating several innovative technologies and design features, including the use of aerogel insulation.

Aerogel insulation is a highly effective insulation material that provides superior thermal performance while being extremely lightweight. Aerogel is made up of 95% air and has a low thermal conductivity, which makes it an excellent insulator. The use of aerogel insulation in the Net-Zero Energy Research House helped to reduce heat loss through the walls and roof, which in turn reduced the energy required for heating and cooling the building. This allowed the building to rely solely on renewable energy sources, such as solar panels, to meet its energy needs. The use of aerogel insulation in the walls and roof of the Net-Zero Energy Research House was a key factor in helping the building achieve its net-zero energy consumption goal. The insulation was applied as a spray-on coating that adhered to the walls and roof, filling all gaps and voids in the structure. The use of aerogel insulation allowed the walls and roof to achieve an R-value of 40, which is twice the insulation value required by building codes in the region.

In addition to providing superior insulation performance, the use of aerogel insulation had several other benefits for the Net-Zero Energy Research House. For example, because aerogel insulation is hydrophobic, it helped to prevent moisture from penetrating the walls and roof, which is important for maintaining the integrity of the insulation and preventing mold growth. Additionally, because aerogel insulation is lightweight, it helped to reduce the overall weight of the structure, which was important for the foundation and overall structural design. One of the key challenges associated with the use of aerogel insulation is the cost. Aerogel insulation is currently more expensive than traditional insulation materials, which can make it difficult for builders and designers to justify its use in their projects. However, the long-term cost savings associated with reduced energy consumption can help to offset the initial cost of aerogel insulation. Additionally, as more builders and designers begin to incorporate aerogel insulation into their projects, economies of scale may help to reduce the cost of the material over time.

Another challenge associated with the use of aerogel insulation is the installation and handling of the material. Because aerogel insulation is extremely lightweight and fragile, it can be difficult to install without damaging the material. Additionally, because the insulation is applied as a spray-on coating, it requires specialized equipment and expertise to ensure proper application. Despite these challenges, the use of aerogel insulation in the Net-Zero Energy Research House demonstrates the significant potential of this material in sustainable building construction. The success of the project has helped to inspire other designers and builders to incorporate similar sustainable technologies and designs into their own projects. The Net-Zero Energy Research House has received several awards and accolades for its innovative use of technologies and designs to achieve net-zero energy consumption. In 2015, the project received the Energy Efficiency Award from the Canadian Home Builders' Association, and in 2016, it was named the Green Building of the Year by the Ottawa Green Building Council. The Net-Zero Energy Research House is a prime example of how aerogel insulation can be used in sustainable building construction to achieve superior thermal performance and reduce energy consumption. Despite the challenges and limitations associated with the installation and handling of aerogel insulation, the benefits it provides in terms of energy efficiency and sustainability make it an attractive option for designers

CHAPTER EIGHT: FUTURE TRENDS IN AEROGEL APPLICATIONS IN SUSTAINABLE BUILDING CONSTRUCTION

Research and Development

Aerogel, with its remarkable properties, has the potential to revolutionize sustainable building construction. Its thermal insulation, acoustic insulation, and fire resistance capabilities make it an ideal material for use in buildings. As the need for sustainable building construction grows, so does the demand for advanced materials such as aerogel. Research and development in this field are essential to further improve the properties of aerogel and to expand its use in sustainable building construction. One area of focus for future research and development in aerogel is the creation of new types of aerogel composites that are specifically tailored to meet the demands of different building applications. Aerogel composites that are optimized for sound insulation or fire resistance could be highly beneficial in reducing energy consumption and increasing safety in buildings. By combining aerogel with other materials such as polymers or metals, researchers can create materials with unique properties that can be optimized for different applications.

The manufacturing process for aerogel is currently energy-intensive and expensive, making it less commercially viable. Researchers are working on developing more cost-effective manufacturing processes for aerogel to reduce costs and increase production efficiency. One method that has shown promise is 3D printing, which has the potential to reduce production costs and provide a more efficient and sustainable method of manufacturing aerogel-based materials.

Improving the mechanical strength and durability of aerogel-based materials is another critical area of research and development. While aerogel is highly effective at insulation, it is relatively fragile and can be prone to damage over time. Researchers are investigating ways to improve the mechanical strength and durability of aerogel-based materials by incorporating nano-reinforcements or improving the manufacturing process to enhance material properties. The potential use of aerogel in transparent insulation materials for windows and skylights is another exciting area of research. Transparent insulation materials that incorporate aerogel could significantly reduce energy consumption by providing natural light and reducing the need for artificial lighting and heating. This type of material would be highly beneficial in buildings with limited access to natural light and where energy consumption is a significant concern.

There is also growing interest in the use of aerogel for energy storage in buildings. Researchers are exploring the possibility of combining aerogel with phase-change materials, which can store and release thermal energy. These materials could be used in buildings to store thermal energy during the day and release it at night, reducing the need for heating and cooling systems. This type of material could be highly beneficial in off-grid or remote areas where traditional energy sources are not available. Finally, aerogel has the potential to significantly reduce carbon emissions in the construction industry. As the demand for sustainable building construction continues to grow, the use of advanced materials such as aerogel will play an increasingly important role in reducing carbon emissions. Continued research and development in this area will be essential in realizing the full potential of aerogel in sustainable building construction. The future of aerogel applications in sustainable building construction is highly promising. Ongoing research and development efforts are focused on improving the performance, durability, and cost-effectiveness of aerogel-based materials. As these efforts continue, it is likely that we will see an increase in the use of aerogel in sustainable building construction, helping to reduce energy consumption and greenhouse gas emissions while improving building performance and comfort.

Large-Scale Implementation

As concerns for environmental sustainability and energy efficiency continue to rise, the demand for innovative building materials and technologies is also increasing. Aerogel, with its exceptional insulation properties and sustainable features, is a promising candidate for the future of sustainable building construction. While aerogel is already being used in various building applications, large-scale implementation of aerogel in construction remains a challenge. This article will discuss the potential future trends in the large-scale implementation of aerogel in sustainable building construction.

One of the major challenges in the large-scale implementation of aerogel is the cost of production. Currently, aerogel is still considered a relatively expensive material due to the high cost of production. However, with the advancements in technology and increased demand, it is expected that the cost of production will decrease over time. Research and development in the production process of aerogel, such as the use of cheaper raw materials and more efficient production techniques, will also help to reduce the cost of aerogel. Another challenge in the large-scale implementation of aerogel is its availability. While aerogel is not a rare material, its production is limited to a few companies. To increase the availability of aerogel, more companies need to invest in the production of aerogel, and more research should be conducted to find alternative methods of production that can produce aerogel at a larger scale. Additionally, efforts should be made to increase the public awareness and demand for sustainable building materials, which will in turn encourage more companies to invest in the production of aerogel.

The installation and handling of aerogel also present challenges in large-scale implementation. Aerogel can be fragile and difficult to handle, which can result in breakage and damage during installation. Therefore, it is important to develop new installation methods and systems that can facilitate the handling and installation of aerogel, making it easier and safer to work with. Furthermore, research and development efforts should be directed towards improving the properties of aerogel, such as its strength, durability, and fire resistance. This will increase its versatility and applicability in different building applications, and further increase its demand in the construction industry.

One of the potential future trends in the large-scale implementation of aerogel is the integration of aerogel into the building envelope system. The building envelope is the outer layer of a building that separates the indoor and outdoor environments, and includes the walls, roof, windows, and doors. By incorporating aerogel into the building envelope, buildings can achieve higher energy efficiency and improved insulation, resulting in reduced energy consumption and greenhouse gas emissions. In addition, aerogel can also be used in the construction of prefabricated building components, such as panels and modules. These prefabricated components can be assembled off-site and transported to the construction site for installation, resulting in reduced construction time, costs, and waste. This method of construction can also improve the quality and consistency of building components, resulting in better-performing buildings.

Another potential future trend is the use of aerogel in the retrofitting of existing buildings. Many existing buildings are energy inefficient and require costly upgrades to meet current energy efficiency standards. Aerogel can be used in the retrofitting of these buildings to improve insulation and energy efficiency, resulting in significant energy savings and reduced greenhouse gas emissions. Finally, aerogel can also be used in the construction of greenhouses and other agricultural structures. Greenhouses require a stable and controlled environment for plant growth, which can be achieved through the use of aerogel insulation. By using aerogel in greenhouses, growers can achieve greater energy efficiency and reduced operating costs, resulting in a more sustainable and profitable agricultural operation. Large-scale implementation of aerogel in sustainable building construction requires a concerted effort from various stakeholders, including building owners, architects, engineers, manufacturers, and policymakers. To achieve this goal, several strategies can be implemented, including:

Cost Reduction: One of the major barriers to large-scale adoption of aerogel in building construction is the high cost of production and installation. To overcome this challenge, researchers and manufacturers need to develop more cost-effective production processes and installation methods. Policymakers can also incentivize the use of aerogel in building construction by offering tax credits, grants, or other financial incentives.

Standardization: As with any new building material, standardization is essential to ensure that aerogel products are safe, effective, and comply with building codes and regulations. Standards and guidelines should be developed to ensure that aerogel materials and products are tested, labeled, and certified for use in sustainable building construction.

Education and Training: Building professionals, including architects, engineers, and contractors, need to be educated on the benefits and applications of aerogel in building construction. Training programs should be developed to teach building professionals about the proper installation and handling of aerogel materials.

Collaboration: Large-scale adoption of aerogel in building construction requires collaboration among various stakeholders, including researchers, manufacturers, building owners, architects, and policymakers. Collaboration can lead to the development of innovative solutions and the sharing of best practices and knowledge.

Large-scale implementation of aerogel in sustainable building construction holds great promise for reducing energy consumption, improving indoor air quality, and promoting sustainability. With continued research and development, cost reductions, standardization, education, and collaboration, aerogel can become a mainstream building material in the future.

CHAPTER NINE: COMPARISON WITH OTHER BUILDING MATERIALS

Comparison with traditional insulation materials

Aerogel insulation offers several advantages over traditional insulation materials, such as fiberglass, cellulose, and foam:

Higher Insulation Value: Aerogel insulation has a much higher insulation value than traditional materials. It has a thermal conductivity rating of 0.013-0.018 W/mK, compared to fiberglass with a rating of 0.04-0.05 W/mK, cellulose with a rating of 0.038 W/mK, and foam with a rating of 0.03-0.05 W/mK. This means that aerogel can provide the same insulation value as traditional materials with a much thinner layer, making it ideal for applications where space is limited.

Fire Resistance: Aerogel is inherently fire-resistant, with a Class A fire rating. Traditional insulation materials are typically treated with fire retardants to achieve a similar rating. However, these chemicals can release toxic gases when burned, posing a health risk to occupants.

Moisture Resistance: Aerogel insulation is hydrophobic, meaning it repels water. This makes it ideal for applications where moisture can be a problem, such as in humid environments or in walls and roofs where moisture can build up.

Durability: Aerogel insulation is extremely durable and can last for decades without degrading or losing its insulating properties. Traditional insulation materials, such as fiberglass, can break down over time, losing their effectiveness.

Environmentally Friendly: Aerogel insulation is made from sustainable materials and has a low environmental impact. It does not contain harmful chemicals, such as formaldehyde, that can be present in traditional insulation materials. Additionally, it is recyclable and can be reused at the end of its lifecycle.

However, there are also some limitations and challenges to using aerogel insulation:

Cost: Aerogel insulation is more expensive than traditional insulation materials, which can be a barrier to its widespread adoption. However, as the technology continues to develop and become more widely used, the cost is expected to decrease.

Availability: Aerogel insulation is not yet widely available, and it can be difficult to find suppliers and contractors who are knowledgeable about its use and installation.

Handling and Installation: Aerogel insulation is fragile and can be difficult to handle and install. It requires specialized equipment and expertise, which can add to the cost of installation.

Limited Applications: While aerogel insulation is suitable for a wide range of applications, it may not be the best choice for all projects. For example, in some cases, traditional insulation materials may be more appropriate or cost-effective.

Despite these challenges, aerogel insulation has the potential to revolutionize the way buildings are insulated and constructed, offering a more sustainable, efficient, and safe solution than traditional insulation materials. As research and development continue to improve the technology and reduce the cost, it is likely that we will see increased adoption of aerogel insulation in sustainable building construction.

Comparison with other sustainable building materials

Aerogel is just one of the many sustainable building materials that are available today. As such, it is important to compare it with other sustainable building materials to determine how it stacks up in terms of performance, cost, and overall sustainability. In this section, we will compare aerogel with other popular sustainable building materials, including cellulose insulation, mineral wool insulation, and spray foam insulation. Cellulose insulation is a popular sustainable building material that is made from recycled paper products. This insulation material is easy to install and has a relatively low cost. It is also an effective insulation material, with an R-value of between 3.5 and 3.8 per inch. However, cellulose insulation is prone to settling over time, which can decrease its effectiveness. It is also not as effective in preventing air infiltration as other insulation materials, such as spray foam insulation.

Mineral wool insulation is another sustainable insulation material that is made from natural rock materials. This insulation material is known for its fire resistance, acoustic properties, and thermal insulation properties. It is also resistant to moisture and pests, making it a durable insulation material. However, mineral wool insulation is not as effective as other insulation materials, such as spray foam insulation, in preventing air infiltration. It is also more expensive than other insulation materials, including aerogel.

Spray foam insulation is a popular sustainable insulation material that is made from a mixture of polyurethane and isocyanate. This insulation material is known for its ability to seal gaps and prevent air infiltration, making it one of the most effective insulation materials available today. It is also an excellent thermal insulation material, with an R-value of between 3.5 and 6.5 per inch. However, spray foam insulation is more expensive than other insulation materials, including aerogel. It is also more difficult to install and may require specialized equipment and training. When compared to these sustainable building materials, aerogel has several advantages. For one, it has the highest R-value of any insulation material available today, making it one of the most effective insulation materials available. It is also lightweight, easy to install, and does not settle over time, making it a durable insulation material. Additionally, aerogel is resistant to fire, moisture, and pests, making it a durable and long-lasting insulation material.

In terms of cost, aerogel is more expensive than some of the other insulation materials available today, including cellulose insulation and mineral wool insulation. However, it is less expensive than spray foam insulation, making it a more cost-effective alternative for those looking for a high-performance insulation material. Additionally, as the technology for producing aerogel continues to improve and production volumes increase, it is expected that the cost of aerogel will continue to decrease, making it an even more viable insulation option for sustainable building construction. While there are several other sustainable building materials available today, aerogel stands out as a high-performance insulation material that is both durable and easy to install. As the technology for producing aerogel continues to improve, it is likely that we will see even more applications of this sustainable building material in the future.

CONCLUSION

Aerogel is a highly promising sustainable building material due to its unique properties such as high thermal insulation, fire resistance, and soundproofing capabilities. Its use in building construction can lead to energy savings, reduced carbon footprint, and long-term cost savings. However, there are also challenges and limitations to its use, including availability, cost, and handling difficulties. Two case studies of aerogel applications in sustainable building construction, NASA's Sustainability Base and the Net-Zero Energy Research House, demonstrate its potential for achieving net-zero energy performance and creating a comfortable indoor environment.

Future trends in aerogel applications in sustainable building construction include further research and development to improve its properties and large-scale implementation in the industry. When compared to traditional insulation materials, aerogel offers superior thermal insulation, but it is more expensive and less readily available. When compared to other sustainable building materials, such as cellulose insulation or foam glass, aerogel offers better thermal insulation and fire resistance, but its cost and availability remain limitations. Aerogel shows great promise as a sustainable building material, and continued research and development can help overcome its limitations and make it more widely available and affordable for the construction industry.

The future outlook for aerogel in sustainable building construction is promising, as it offers numerous benefits such as energy efficiency, reduced carbon footprint, and long-term cost savings. The development and application of aerogel in the building industry have the potential to revolutionize the way we design and construct buildings. One major future trend in aerogel application is its continued research and development. There is ongoing research to improve its properties, such as increasing its mechanical strength and reducing its cost. As these improvements are made, it is likely that aerogel will become more widely used in the construction industry.

Another trend is the increased demand for sustainable building materials due to growing concerns about climate change and the need for more energy-efficient buildings. The adoption of green building practices has become more widespread, and as such, there is a growing market for sustainable building materials. As aerogel becomes more widely available and affordable, it is likely to become a popular choice for building insulation and other applications.

There is an increasing focus on net-zero energy buildings, which are designed to produce as much energy as they consume. Aerogel's superior thermal insulation properties make it an ideal material for achieving net-zero energy performance. As the demand for net-zero energy buildings increases, the use of aerogel in construction is likely to follow suit.

In conclusion, aerogel has shown great potential as a sustainable building material with various applications in the construction industry. Its unique properties, such as its high thermal insulation, fire resistance, and sound absorption capabilities, make it a desirable alternative to traditional insulation materials. Additionally, aerogel's long-term cost savings, health and safety benefits, and reduced carbon footprint contribute to its appeal as a sustainable building material. However, there are still some challenges and limitations that need to be addressed before aerogel can become a widely used material in sustainable building construction. These include the high cost and limited availability of aerogel, as well as the challenges associated with handling and installation.

Despite these challenges, there is a growing trend towards research and development of aerogel-based products and systems, as well as increasing large-scale implementation in sustainable building projects. This indicates a promising future for aerogel in the construction industry, as it continues to evolve and improve as a sustainable building material. Ultimately, the use of aerogel in sustainable building construction is just one piece of the larger puzzle in the pursuit of a more sustainable and environmentally conscious future. As the world faces increasing environmental challenges, it is important for the construction industry to continue to innovate and explore new materials and technologies that can contribute to a more sustainable future.

ABOUT THE AUTHOR

Steven Smith

Steven Smith is a respected expert that holds a Doctorate in Construction Management. His dedication to sustainability and passion for educating others make him a true leader in the field.

BOOKS BY THIS AUTHOR

Building Maintenance Guidelines: A Complete Manual

This book is an essential resource for anyone responsible for maintaining and preserving the integrity of a building. It covers several aspects of building maintenance, from electrical systems and HVAC systems to roofing, plumbing, and structural components. It provides clear, step-by-step instructions on how to perform routine maintenance tasks. It also includes information on how to identify potential problems, such as water damage, mold growth, and insect infestations, and provides guidance on how to address these issues. In addition to its practical information, the book also includes important information on energy efficiency and sustainability.

With its clear, easy-to-follow language, the book is an invaluable resource for anyone looking to keep their building in optimal condition.

Dictionary Of Construction And Civil Engineering Terminologies: A Reference For Students, Practitioners, Academics, And Homeowners

This book is a comprehensive reference guide for anyone involved in the construction and civil engineering fields. It is a dictionary containing an extensive range of terminologies and technical jargon used in the industry. This makes it an essential tool for students, practitioners, academics, and even homeowners who want to take their understanding of the industry to the next level.

The book covers a vast range of topics, including architectural design, building construction, structural engineering, environmental engineering, and much more. It provides clear, concise, and accurate explanations for each concept and technical language.

With thousands of terms, the book will help you navigate the technical language and complexities of the field. Are you a student, practitioner, academic, or homeowner looking to expand your knowledge and understanding of the construction and civil engineering industry? Look no further than this book.

Construction Management Fundamentals: A Handbook For Construction Students, Academics, And Practitioners

Are you seeking a comprehensive guide to mastering the fundamentals of construction management? Look no further than this book! This handbook offers a clear and concise overview of the principles, practices, and techniques of construction management. It covers the concepts essential for a successful career in construction management, including project planning, scheduling, and construction contracts. The book is designed to be an invaluable resource for construction students, academics, and practitioners alike.

Whether you're a seasoned professional looking to refresh your skills or a student just starting out in the field, this handbook is an indispensable resource that will help you excel in your career.

www.ingramcontent.com/pod-product-compliance
Lightning Source LLC
Chambersburg PA
CBHW071028220526
45467CB00004B/1568